LUFTWAFFE AT WAR

Stukas over the Steppe

While armourers work on the main bomb beneath the fuselage, other members of the Stab./St.G. 77 relax in the dappled sunlight of nearby woods while this 'Dora-3' is made ready for her next mission. Note that her pilot's front windscreen has the normal addition of a flat, 50mm armoured glass windshield with extra bracing, behind which is located the Revi C-12-C bombsight. For further protection the seat-mounted headrest was fully armoured as was the overturn structure immediately abaft it. (*James V. Crow*)

Stukas over the Steppe

Peter C. Smith

Pen & Sword
AVIATION

Stukas over the Steppe

A Greenhill Book
First published in 1999 by Greenhill Books,
Lionel Leventhal Limited
www.greenhillbooks.com

This edition published in 2015 by
PEN & SWORD AVIATION
An imprint of
Pen & Sword Books Ltd
47 Church Street
Barnsley
South Yorkshire
S70 2AS

Copyright © Lionel Leventhal Limited, 1999

ISBN: 978-1-84832-801-3

CIP data records for this title are available from
the British Library

Designed by DAG Publications Ltd
Design by David Gibbons
Layout by Anthony A. Evans

Printed and bound in Malta by Gutenberg Press Ltd

Pen & Sword Books Ltd incorporates the Imprints
of Aviation, Atlas, Family History, Fiction, Maritime,
Military, Discovery, Politics, History, Archaeology,
Select, Wharncliffe Local History, Wharncliffe True
Crime, Military Classics, Wharncliffe Transport,
Leo Cooper, The Praetorian Press, Remember When,
Seaforth Publishing and Frontline Publishing.

For a complete list of Pen & Sword titles
please contact
PEN & SWORD BOOKS LIMITED
47 Church Street, Barnsley, South Yorkshire,
S70 2AS, England
E-mail: enquiries@pen-and-sword.co.uk
Website: www.pen-and-sword.co.uk

LUFTWAFFE AT WAR
STUKAS OVER THE STEPPE
THE BLITZKRIEG IN THE EAST
1941–1945

The launching of the German attack on the Soviet Union on 22 June 1941, Operation BARBAROSSA, was the ultimate and logical climax to the road that the Führer Adolf Hitler had taken since he came to power in 1933. The Communist had long been the ideological enemy of the Nazi state, and that state had been built with the avowed intent to create *Lebensraum* ('living space') for the German people at the expense of the Slavs of central and eastern Europe. The fact that the lands he coveted were under the domination of his most hated opponents only lent inevitability to Hitler's fateful decision. Marshal Joseph Stalin's aggression – the attack on Finland, the occupation of Estonia, Latvia and Lithuania, the annexation of parts of Romania – drew the Soviet Union closer to Germany's eastern borders and added fuel to the fire of deep resentment that Hitler felt towards Russia, making the final confrontation a certainty.

Neither was there much doubt about the methods by which the German armed forces meant to take on and defeat the huge military machine that Stalin had built up over the years. Far outnumbered in everything – infantry divisions, artillery, tanks and aircraft – the Germans still took on the task with supreme confidence. This was due to the fact that their military and aviation strength had been tried and tested in full combat. In *Blitzkrieg* operations against Poland, Norway, the Netherlands, Belgium, France, Yugoslavia and Greece, the highly mobile and fast-moving combination of armoured column thrusts, supported by the 'flying artillery' provided by the dive-bombing Junkers Ju 87 *Stuka* (Sturzkampfflugzeug – dive-bomber) units, had proved irresistible.

In campaign after campaign the main power of the opposing armies had been broken by the *Stuka*s and crushed by the *Panzers,* with conventional forces mopping-up the dazed and disorganised remains of once-proud armies in their wake. The *Luftwaffe* fully co-operated with the *Wehrmacht* and their liaison was years in advance of any other nation. Army support, in *Luftwaffe* terms, was wholehearted and unreserved. The concentration of all forces on one vital spot, the *Schwerpunkt,* broke initial resistance in the form of fortresses, strong-points and concentrations of troops; then, once the breakthrough was achieved, the tank columns raced off, surrounding and enveloping the less mobile armies of their opponents. Never given time to reorganise, always one or two steps behind, more conventional armies were swept away.

Whenever a stand was made, and forces tried to rally, or a counter-attack was mounted against the German advance, the *Stuka*s were called in and effectively smashed the enemy resistance before it could gather momentum. The same tried-and-tested principles were to be employed again in BARBAROSSA, but against a far tougher opponent.

Although Stalin's purges of the Soviet Army High Command in the late 1930s had resulted in the loss of thousands of competent officers, and the Finnish debacle in the winter of 1939/40 had shown up huge failings in Soviet military thinking and tactics, the colossal Soviet war machine was equipped with the world's toughest tanks, and thousands of them, and an inexhaustible supply of manpower, however badly trained. Most important of all, however, Russia had vast tracts of land, the huge empty steppe that stretched for hundreds of miles eastward. Trading space for time was an old Russian war tactic and had brought Napoleon's dreams to a terrible end in the harsh winter of

1812. It was ultimately to do the same to Hitler's towering ambitions. Twenty-two miles of salt water had saved Great Britain from going the same way as her allies in the summer of 1940, and the endless steppe was to serve Stalin in the same way.

Under *Generalfeldmarschall* Albert Kesselring's *Luftflotte* 2, supporting Army Group Centre, were concentrated the *Stuka*s of St.G. 77 of *General der Flieger* Bruno Loerzer's II.*Fliegerkorps* and *Generalfeldmarschall* Wolfram *Freiherr* von Richthofen's VIII.*Fliegerkorps*, with the St.G. 1 and the St.G. 2. Under *Luftflotte* 5, commanded by *General der Flieger* Hans-Jürgen Stumpff, was the IV.(St.)/LG. 1. In all a total of 424 *Stuka* dive-bombers were available to take on the Soviet Union – only ten more than had been available on the opening day of the attack on France and the Low Countries just over a year before.

Initially, however, *Blitzkrieg* worked well, in fact better than ever. In the first week of the fighting an incredible 4017 Soviet aircraft were destroyed, mainly on their airfields, for the loss of just 150 German aircraft. In the great encirclement battles of 1941, hundreds of thousands of Russian soldiers were trapped, defeated and taken prisoner. Unprecedented numbers of tanks, vehicles and artillery pieces were captured as the dive-bombers performed their role and spearheaded the drive eastward.

One thing proved to be a powerful portent for the future, however. The vast size of the Eastern Front, a thousand-mile long battlefield from the snowy wastes of Murmansk in the north to the Black Sea littoral in the south, ensured that the *Schwerpunkt* principle was now essential. There were just not enough *Stuka*s to achieve local superiority at all points down the blazing line. Instead the *Stukagruppen* were switched from front to front, moving airfields almost daily as the objectives switched and opportunist advantage was taken to round up yet more bewildered and dazed enemy armies.

Such intensity of operation would have been impossible for most air forces to sustain for any length of time, but the *Luftwaffe* had been designed to operate in such a way; rapid moves of their frontline units had been a feature of all the previous campaigns. It had the ability to switch dive-bombers from the Channel coast facing England to Romanian airfields, in order to smash Yugoslavia and Greece, and then on to Polish and other Romanian bases in readiness to deal with Russia. All this had been achieved smoothly and efficiently across difficult terrain.

But the very vastness of the new battlefield brought about its own attrition. Soviet Russia, despite its appalling losses, did not cave in like Belgium or France, and although it learned its lessons slowly, it learned them well. The Germans failed to reach all their main objectives; Murmansk, Leningrad, and Moscow all were threatened, almost seemed within grasp, but, finally, eluded the Germans. Then the Soviets' other great allies stepped in and saved the day: Generals Mud and Winter arrived in swift succession and brought fast-moving tank columns to a grinding halt.

Just what conditions were like for *Stuka* units is hard for us to imagine. Used to short campaigns in moderate climates, the huge range in weather conditions encountered on the immense Eastern Front varied from one extreme to the other. In the far north those *Stukagruppen* assigned to the advance on Murmansk found themselves in treeless, isolated tundra. There were no roads – not even paths – just a desolation of rock and scree. For four brief months in the summer this region of icy permafrost, scored by blizzards and with temperatures of minus 50°C, turned into a mosquito infested swamp criss-crossed by lakes, rivers and streams which made the sixty miles the Germans had to advance a nightmare. This contrasted with conditions found at the southern end of the front, where the steppe rolled endlessly and, in high summer, turned into a red and brown desert of swirling dust which, when whipped up by high winds, penetrated everything. By autumn this barren brown plain was quickly turned into a quagmire of mud into which men, horses and light vehicles could disappear whilst heavy vehicles and tanks became bogged down.

An early winter, and the winters of 1941/42 and 1942/43 were exceptionally hard, even for Russia, transformed the mud into frozen hard earth, enabling quick movement, but as the temperatures continued to plummet this presented the *Stukagruppen* with new nightmares. These first came to light in the final advance on Moscow in 1941 when the temperature suddenly dropped. Aircraft serviceability fell appreciably in these conditions and, although the strong and rugged *Stuka*s were better able to cope than most aircraft, even they were badly affected.

With the first snow came problems of clearing the airfields so that the dive-bombers could get airborne at all, but it was the sharp drop in temperatures that really caused the problems. One night minus 40°C was recorded and the engines would not start as the lubricating oil was frozen solid, and the cold was so intense that even the wing machine-guns jammed. The campaign had been planned as an intense six week operation, over by September at the latest, but December came and with it the appalling weather. The *Luftwaffe* found itself unprepared.

Temperatures continued to plunge, minus 50°C being recorded. It was not just the immediate affect on the aircraft that caused problems, but also the difficulties with replenishment. The Germans were now operating at the end of a long supply route from Europe and everything – food, fuel, ammunition, ordnance, spare parts and basic essentials – had to reach them via the already congested road and rail systems. The bad weather also brought these to a grinding halt. Thus the Ju 87s, if they could be started at all, stood immobilised for lack of fuel, ammunition and oil.

Then there was the effect on personnel, all of whom lacked basic winter clothing for the first months, with hardships and loss of efficiency. Aircrew had to fly four or five sorties each and every day and their 'Blackmen' ground crews (named for their black overalls) had to stay awake overnight in the intense cold and try to keep the engines running and prevent them from freezing. Morale was adversely affected and Hans-Ulrich Rudel described conditions at Rhew, near the source of the Volga, during the first winter thus:

'Engines no longer start, everything is frozen stiff, no hydraulic apparatus functions, to rely on technical instruments is suicide. There is no starting our engines in the early morning at these temperatures although we keep them covered up with straw mats and blankets. The mechanics are often out in the open all night long, warming up the engines at intervals of half an hour in order to make sure of their starting when we take off. Many cases of frostbite are due to spending these bitterly cold nights looking after the engines.'*

As December gave way to January, things, if anything, became worse, so that even more intense cold increased the difficulties of making the *Stuka*s serviceable for the next day's missions. This was at a period when the Soviet armies were counter-attacking all along the front, throwing back many units, cutting off others and isolating them in small pockets. It was vital for the *Stuka*s to operate in the face of these attacks if the line was to be stabilised, but it became increasingly difficult. In the end some engineers resorted to desperate measures and Rudel recorded seeing mechanics trying to warm up engines using naked flames. One of them told him, 'They'll either start now or be burnt to a cinder. If they won't [start] they are no use to us anyway.'

In the end a rudimentary solution was found. A makeshift portable heater was made using a petrol can as a tin oven, with a chimney fitted with a cowl to prevent sparks flying directly upwards. This heater was then wheeled beneath the *Stuka*'s engine, a fire was lit and the chimney was pointed at the primer pump until it thawed out. Such methods were crude but they worked. The official solution later arrived at the front – beautifully constructed heat carriers, made specially in Germany. Unfortunately, before they could work they had to be started up too as their own tiny motors were equally frozen solid.

Unprepared for the harsh weather (they had expected that the campaign would be over before winter set in), the Germans halted all along the line, and then the Russian armies went over to the counter-offensive. Now a new task fell to the *Stuka*s. Rather than leading their own armoured thrusts, they had the onerous task of trying to stop the enemy ones. Flying up to five combat missions a day in appalling conditions they helped stabilise the front, but at great cost.

In 1942 more limited objectives were undertaken as the Germans once more went on the offensive. This time the thrusts were south-west to secure the bread basket of the Ukraine and the oil-rich Caucasus. Again the *Stuka*s paved the way, and, against the strongest fortifications in the world, they blasted a path that led to the fall of Sevastopol and the conquest of the Crimea. But again the Germans overreached themselves when they tried to gain the far bank of the Volga.

*Hans-Ulrich Rudel, *Stuka Pilot*, Euphorion Books, Dublin, 1952

Instead of by-passing and enveloping the city of Stalingrad they tried to assault it head-on. They were sucked into a street-by-street and house-by-house slugging match and bled white. Then, with winter looming again, the Soviets again counter-attacked on both flanks and only dedicated work by the *Stuka*s in their ground-attack role temporarily stemmed the tide.

The German armies attacked one final time, in the summer of 1943, at the Kursk salient. The objective was now the very limited one of wearing down the Soviet army in the hope that it would eventually accept a stalemate situation. There was no chance of success in this hope, nor of an actual victory, despite initial progress. The enemy had been fully forewarned of the offensive and was ready. Kursk in 1943 had no more succeeded in breaking Russian will than Verdun had for the French in 1916.

From then onward the *Stuka*s fought a dour, defensive war. Overwhelmed by sheer weight of numbers, and with the British and Americans finally joining in the land war to the rear, firstly in Italy and then in France, the Junkers Ju 87 adapted itself continually, becoming a tank-buster with special 37mm armour-piercing cannon; carrying specialised bomb-loads like the SD-2 'butterfly' bomb used as a cluster weapon against the masses of Soviet infantry; becoming a night bomber; and attacking selected targets on all fronts with some success until the very last day of the war.

The vital need to introduce an aerial weapon to counter the masses of T-34 tanks was seen earlier in the campaign in the east. Although the *Stuka*s had been successful enough against tanks in France and the Mediterranean, the new enemy had thousands upon thousands of them. Bombs did not require a direct hit to disable a tank – a near miss was usually enough to strip off the tracks and immobilise one – and judiciously placed machine-gun bullets had proved effective against lightly armoured western tanks. But the Soviet T-34, and the even heavier 'Stalin' tanks, required a more powerful antidote.

After testing and trials at the *Luftwaffe* research centre at Rechlin, an answer was found in the adoption of the 37mm Flak 18 (*Bord-kanone* 3.7cm) cannon. With an overall weight of 270 kg (600 lbs), this weapon had its origins in a World War I design for an anti-aircraft gun. Known as the Fla 18/36/37, the gun was first used in the anti-aircraft units of the *Luftwaffe* as early as 1933. It did not prove a success in this role, but its ballistics made it an excellent candidate for the penetration of tank armour. The original electric sighting mechanism was unsuitable for aerial warfare but a new pneumatic system was substituted and the weapon seemed viable.

With a length of 3620mm, this powerful gun had a muzzle velocity of 855 metres (2820 feet) per second and could fire a 1.4 kg (3 lb) *Minengranatpatrone 18* tungsten-cored armour piercing explosive shell. Tests proved that this missile could pierce 58mm thick tank armour at a 60° angle at a range of 100 metres. Later developments meant that the weapon could penetrate twice this thickness of tank protection. Five rounds of ammunition were carried in two clips which were fed horizontally via a metal loading tray with a downward hinged metal cap. One gun was mounted on each wing and the actual firing mechanism was shielded aerodynamically in a metal pod along with the weapon's hydraulic heater.

The first trials on the *Stuka* took place on a Ju 87D-1 (*Werk Nr.* 2552) in December 1942. After ground test firings proved satisfactory, aerial testing followed with veteran pilots with a proven anti-tank record being drafted in from the front to take part. Following this an experimental unit was set up, the *Versuchsverband für Panzerkampfung*, under the command of *Oberstleutnant* Otto Weiss. The unit went into action from 18 March 1943 and soon each *Stukagruppe* included at least one *Kanonenvogelstaffel*. The *Stuka* thus took on a whole new career as a tank-buster *par excellence*.

In the end several of the last formations flying the *Stuka* on the Eastern Front flew their aircraft to airfields in the western zone of influence and surrendered themselves as fully-intact units. From day one of World War II to the final act, the Junkers Ju 87 had fought with great distinction in the front line and become one of the most feared and famous warplanes of all time.

Right: Seen from below in classic pose this *Stuka*, with a light spinner, carries the early 'standard' bomb load of one centrally carried 551 lb bomb on the swing-arm crutch and two 50 lb bombs on underwing bomb racks under each wing. As the war in the east progressed the combinations of weaponry carried diversified enormously. The addition of the percussion detonator to the SC-250 bomb allowed it to detonate above ground, increasing the blast effect on the masses of Soviet infantry. Other anti-personnel weapons carried by the Ju 87 included the 'butterfly' bomb carried in the AB 250 Light Weapons Dispenser, and three extra machine-guns each side, for strafing dense formations of infantry, were carried under the wings in the WB 81A/B 'watering-can'. (*Archiv Von Lutz*)

Above: With bare-chested ground crew and a pilot with his sleeves rolled up, these Junkers Ju 87Ds stand ready by the camouflaged dispersal area close to the trees surrounding this forward airfield in the summer of 1942. The unit is VII./St.G. 1, and they carry the blue helmet and red anchor emblem of III./St.G. 1. This denoted their origins as the first 'Navalised' *Stuka*s, the dive-bomber units originally formed to operate from the aircraft carrier *Graf Zeppelin*. Originally designated 4./Trägergruppe 186, the unit was, with the first cancellation of the project, incorporated into the *Luftwaffe* mainstream as from July, 1940. Now operating a long way from the water, the almost featureless sea of the Russian steppe must have seemed as endless and boundless as the ocean they had originally been formed to patrol. (*Archiv Von Lutz*)

Above: A Junkers Ju 87B at rest on a Balkan airfield prior to a secret and lightning re-location to forward area airstrips close to the Soviet Union in readiness for Operation BARBAROSSA. Although the decision to attack in the east came as some surprise to the *Luftwaffe*, the efficiency of its organisation meant that this complicated redeployment from Greece via Romania and Germany to Poland, across one of the more backward areas of Europe, was successfully undertaken in the startling time of three weeks. This built-in potential for flexibility was as impressive a part of the *Stuka*'s versatility as its accuracy and ruggedness. (*Archiv Von Lutz*)

Below: The pattern of the Eastern Front in its early months. Spread across a captured Soviet airfield in the middle of the vast and featureless Russian steppe, the Junkers Ju 87B-1s of a *Stukagruppe* stand ready for mass take-off to their assigned target. Up to four dive-bombing missions a day was typical for *Stuka* aircrew. Moving forward almost daily in the early months of the campaign and again spearheading the Panzer formations as they outflanked and surrounded thousands of the demoralised Soviet enemy, the *Stuka*s were at the forefront of the drive on Moscow. In the foreground is a petrol bowser, with the unit's transport behind it. In the centre are two captured Soviet aircraft, two Polikarpov I-16 fighters with red star markings on their wings, hastily abandoned in the face of the onrushing *Blitzkrieg*. In the background, a German motorised column streams across the horizon in close pursuit of the ever-advancing Panzers. (*Archiv Von Lutz*)

Above: The pattern of the Eastern Front in its later years. With dappled 'snow camouflage' this Ju 87D is being serviced on a makeshift airstrip from which the snow has just been cleared. With the coming of the second winter to the Front the *Stuka* found itself fighting a somewhat different war. Still at the forefront of the immense battle, and still flying non-stop missions daily, and in atrocious conditions that tested men and machines to their limit, now the *Stuka-geschwader* flew their 'flying artillery' sorties in support of their hard-pressed, and always outnumbered, comrades on the ground as mobile dams against the flooding tide of Soviet manpower that threatened to engulf them. Again and again their targets were the enemy tanks of the armoured columns that pressed against the German defences from all sides, and only the devoted efforts of the *Stuka*s prevented breakthroughs on innumerable occasions. (*Archiv Von Lutz*)

Left: The Ju 87D-1s of VII./St.G. 1 lined up on this forward airfield show the leaner and meaner profile on the improved *Stuka*, with re-contoured cowling, to give a much more streamlined configuration, and smaller oil cooler intake and underwing radiators. The canopy was also rounded-off and incorporated the new twin MG 81Z machine-gun and the GSL-K 81 armoured turret. The dictates of the Russian weather have also made themselves apparent in this photograph, with the wheel spats being removed to help cope with either the dust and chunks of hard, dry earth in the summer, or the dire mud of the steppes in autumn which transformed the dusty plain into a sea of clogging glue. Some *Stuka*s had their entire undercarriage covers taken off to cope with these conditions. (*Archiv Von Lutz*)

Cranking up the inertia-starter of a Ju 87B-1 'Bertha' (carrying the number '6' on her nose and white discs on each wheel spat), to kick over her 1100 hp Junkers Jumo 211 A engine. Note the permanently open crank port; on the later B-2 this would be shielded by a hinged cover plate. Also note the smooth after cowling and oblong exhaust ports and after vent, and, above, the auxiliary air intake with the large cover of the 31 litre oil tank just ahead of the cockpit. The three-man team of 'Blackmen' give a good indication of the actual size of the *Stuka*. The red triangle beneath the closed circular oil filling point aperture indicates the specification to be used. (*Archiv Von Lutz*)

A Junkers Ju 87R 'Richard' under camouflage netting at its home base of Bonn/Hungelar airfield prior to the Russian campaign. Many of these units had only recently seen hard combat service in the *Blitzkrieg*, which crushed Yugoslavia and Greece in the space of a few weeks, and had inflicted a heavy defeat on the Royal Navy off the sunlit waters south of Crete. Now, hastily transferred back to home bases, they were being refitted and readied before making another fast and secret redeployment to forward airfields in occupied Poland. This was in readiness for the launch of Operation BARBAROSSA and what was expected to be the final settling of accounts with the Communist enemy, still nominally an ally. The *Unteroffizier* checking off his list of tasks already proudly wears his Knights Cross for previous accomplishment in action. (*Archiv Von Lutz*)

Left: In front of their red-spinnered *Stuka*, *Hauptmann* Anton Keil (second from left), the *Kommandeur* of II./St.G. 1, goes over the target map with his pilots prior to a strike. Keil had a long and distinguished career in the *Luftwaffe* from 1934 onwards, and was one of the first *Stuka* pilots, and became *Staffelkapitän* of the III./St.G. 162 in March 1937. After taking part in the Polish campaign he led his new command with great distinction in the French campaign and in the early stages of the Battle of Britain. He was to become one of the first *Stuka* casualties of the Russian war when, on 29 August 1941, after conducting an attack east of Skalova, his aircraft was forced to make an emergency landing but turned over in a morass of swamp severely injuring both Keil and his radio operator, *Feldwebel* Knof (extreme left). They were both burnt to death when the upturned aircraft caught fire. Keil and his pilots are wearing the standard issue summer flying uniform of tan coloured flying suit, with diagonal zip fastener. Keil wears the standard *Luftwaffe* peaked cap, while his pilots wear their side caps. (*Signal via author's collection*)

Above: Over the Black Sea coastline, near the Crimea, this Junkers Ju 87D-1 of 7. *Staffel*, III. *Gruppe/Stukageschwader 77* is seen with full bomb load and on her way for yet another strike at the Soviet fortress of Sevastopol in June 1942. This unit had only re-equipped with the 'Dora' in March 1942, during a period of temporary retirement from the front at its main base of Boblingen, near Stuttgart, the 6./St.G. 77 being the first to so re-equip. By May 1942, they were based back in the Crimea, at Sarabusy-South airstrip, closed to Sinferopol, in readiness to lead the final assault on the surrounded but declared 'impregnable' Soviet fortress and fort. The *Stuka*s made their first assaults on 7 May and continued for a week before victory was achieved. (*Archiv Von Lutz*)

Above: A *Stuka* over Stalingrad. A symbolic sunset over the Volga silhouettes a homeward-bound 'Dora' after yet another pin-point attack against stubborn defenders amid the rubble. The setting sun also marks the final and furthermost achievement of the Junkers Ju 87 against the Soviet Union, for, with the enemy counter-attack just weeks away, never again would the high-tide of the *Stuka*/Panzer combination reach further east. From this point on the Junkers dive-bombers and their crews were destined to fight a bitter defensive war in an heroic, but ultimately vain, attempt to stem the Soviet advance. (*Signal via author's collection*)

Right: The last massed attack in formation of the veteran 'Berthas' of the III./St.G. 77, before their replacement by the 'Dora' in July 1944. With the failure of their intended replacement dive-bomber, the Messerschmitt Me 210, the Junkers Ju 87D had been forced to soldier on and indeed found a new lease of life with the tank-busting *Kanonenvogel* Ju 87G 'Gustav', which was World War II's equivalent of today's battlefield tank destroyer, the Fairchild Republic A-10 Thunderbolt-II. Like the Thunderbolt the *Stuka* was slow but functional; like the Thunderbolt, the *Stuka* was dubbed ugly; but, also like the Thunderbolt, the *Stuka* was supremely pre-eminent at its job. As these new *Stuka*s began to join the units, the *Sturzkampf* units were re-dubbed *Schlacht* (ground attack) units on 10 October 1943. Some units flew *Stuka*s on daylight missions right through to the last day of the war. But, from the summer of 1944, through the last ten months of the war, they were gradually replaced in frontline units by the Focke Wulf Fw 190 and the very last *Stuka*s to be produced left the Bremen works in July 1944. As they were replaced, many Ju 87Ds were re-equipped as night-attack (*Nachtschlacht*) aircraft and also continued the struggle. (*Signal via author's collection*)

Above: A Ju 87B-1 of the I./St.G. 1, seen here at Elbing airfield in East Prussia just prior to the outbreak of war, with Wilhelm Busch's cartoon emblem 'Hans Huckebein' under the windscreen. The raven's body and head were black, its bill yellow. The smooth rear cowling and old-style exhausts indicate that this is an early production model. (*Richard Chapman*)

Below: The angled white strip on the rudder of these Ju 87B-2s, here seen over Germany early in 1941, denote the elite unit IV.(St.)/LG 1 commanded by *Hauptmann* Bern von Brauchitsch. The *Lehrgeschwader* (teaching and development wing) was originally formed to test fly new ideas and equipment as well as develop new fighting techniques. In the *Luftwaffe*'s way of doing things this by no means precluded their employment in the full combat role. The IV. *Gruppe* consisted of the 10., the 11. and the 12. *Staffeln* and fought in the front line in Poland, France, Finland and north Russia before being re-designated as the I./St.G. in February 1942. It continued to serve on the north Russian front under the command, from 23 June 1942, of *Leutnant* Hans-Karl Stepp. (*Archiv von Lutz*)

Left: The rear airfield of the III./St.G. 77 at Biala, Poland, on the eve of the Russian campaign. This *Stuka* unit was based here between 20 and 24 June 1941. Here the men, in relaxed mood, take a short break for a meal whilst preparing their Ju 87Bs for the action to come (from left to right): first class mechanic Hubert Mohs (standing); unknown mechanic (sitting); parachute mechanic Karl Kloth; unknown mechanic; first class mechanic Erwin Scherre; chief foreman mechanic Franz Scharre; artificer Alexei Immerreich; and three unknown mechanics. (*Luibe via Sellhorn Archiv*)

Below: A Junkers Ju 87B-2 sits at Rovaniemi airfield in Finland in June 1941, awaiting the 'go' for the launching of the northern wing of the great assault on the Soviet Union. Note that lavish cockpit armour protection is now installed for the rear-gunner with the fitting of the U-3 conversion, which comprised grey-painted angled armour plating extended up both rear corners of the gunner's windows. The overturn bulkhead between the rear and forward cockpits is also fully armour clad. The circular rotating MG 15 gun mount can also be clearly seen. (*Onni Komulainen via Aviation Museum of Central Finland*)

Right: This bomb, mounted on the swing-crutch of a Bertha-2 of the St.G. 77, is inscribed 'Happy Easter 1941, Romania'. The actuating rods of the trailing-wing flaps can be clearly seen in this photograph, giving the distinctive 'Junkers wing'. (*Wilhelm Landau via James V. Crow*)

Below: Day one of BARBAROSSA and the *Stuka*s head eastward to help deliver perhaps the most successful air strike of all time against the Soviet Air Force. Due to the demands of the Balkan campaigns, only 290 of the 310 assigned *Stuka*s were actually ready to take part in these first attacks. (*via Ken Merrick*)

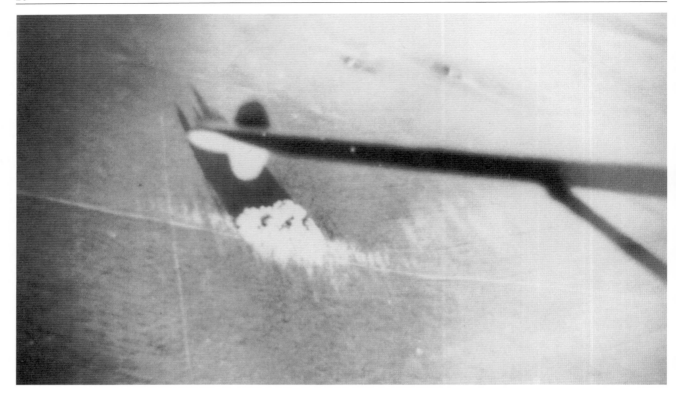

Opposite page, top: An ideal *Stuka* target – the dive-bombers catch a Soviet column in full retreat on the open steppe in September 1941, and punish it severely. In this photograph, taken under the tailplane of a *Stuka* pulling up after its strafing run, Soviet motor transport can be seen exploding under a hail of fire from the following aircraft. (*Frau Orthofer*)

Left: A Ju 87 'Bertha-2' of the IV./St.G 77, coded S2+HM, is seen over the Soviet Union in 1941. Two of her aileron mass balances stand out clearly in this photograph as does the rear-gunner's open vent window. (*Archiv von Lutz*)

Above: The dive-bombing of Junkers Ju 87 was as effective as ever when Soviet artillery columns were caught on the move and summarily dealt with in fine style, as this photograph of the aftermath of an assault by the I./St.G. 77 shows. (*Dr. Armbrust via Sellhorn Archiv*)

Above: The spoils of war. A Soviet I-17 biplane armed with small ground-attack rockets, captured intact during the over-running of Prushany airfield by the German advance in June 1941. The incredible numbers of enemy aircraft destroyed on the ground, shot down or captured was so vast that initially, even the *Luftwaffe* High Command refused to believe the figures. But, after some very careful checking (and the *Luftwaffe* was far more scrupulous at this than the Allied Air Forces ever were), the agreed total came to a staggering 4017 for the loss of only 150 *Luftwaffe* machines, the greatest air victory of all time. (*Maahs via Sellhorn Archiv*)

Below: A Junkers Ju 87B-2, coded S1+GH, of the VII./St.G. 77 on a rough grass landing strip in the Soviet Union during the initial drive forward during the summer of 1941. For the opening of Operation BARBAROSSA weather conditions were ideal and the flower-strewn steppe was like a welcome carpet, but within a few short months it turned into a hellish mire. St.G. 77 initially deployed 177 *Stuka*s as part of II.*Fliegerkorps* and spearheaded Army Group Two's drive across the Bug River. (*Archiv von Lutz*)

Above: A T-34 after the *Stuka*s had dealt with it. A near-miss has brought this tank to a halt and disabled it. The hatch cover is open which indicates that at least one of its crew survived to escape but neither this tank, nor the two equally shattered ones behind it, made any further contribution to the war. (*Dr. Armbrust via Sellhorn Archiv*)

Below: The *Stuka* was a tough aircraft but had its limits. This Junkers Ju 87R-2, coded L1+BU, *Werk Nr.* 5689, of 10.(St.)/LG 1 made a crash landing in dense woodland near Kemijarvi, due east of Rovaniemi, Finland. The wreckage of the aircraft's nose has exposed the oil-cooler saddle tank mounted atop the engine. (*Onni Komulainen via Aviation Museum of Central Finland*)

Above: A battleship becomes a target. At Kronstadt naval base, near Leningrad, lurked the Soviet Baltic fleet with two battleships and a host of lesser vessels. In a series of heavy dive-bombing attacks, mounted between 21 and 24 September 1941, one of the larger vessels, the *Marat,* had her bows blown off. The other battleship, *Oktyabrskaya Revolutsia* (23,000 tons), was heavily damaged by a series of direct hits from the *Stuka*s of the I. and III./St.G. 2 'Immelmann' under *Major* Oskar Dinort. Here the *Oktyabrskaya Revolutsia* is seen from the attacking dive-bombers, surrounded by the splashes of near-misses and with the smoke of one direct hit forward being joined by a second direct hit on the port side aft. In all, she was hit by six medium-sized bombs. She had to be towed to Leningrad to be repaired. In further attacks on 4 April, 1942, an assault by sixty-two Ju 87s from the III./St.G. 1 and the II./St.G. 2 claimed to have hit her again four more times. The *Stuka*s carried special 1000 kg bombs, designed to penetrate battleship armour, on these missions. (*Franz Selinger*)

Opposite page, top: The Soviet heavy cruiser *Kirov* (8107 tons), with the Chief of Light Naval Forces, Rear-Admiral Drozd, aboard, is shown here under *Stuka* attack at Kronstadt naval base by the III./St.G. 2 under *Hauptmann* Ernst-Siegfried Steen on his 301st *Stuka* mission, 23 September 1941. During his dive Steen's *Stuka* received a direct hit in the elevator. Knowing his aircraft was doomed, the pilot tried to crash into the cruiser but, because of the damage, just missed, and his aircraft crashed into the sea.

Before impact he managed to release his 1000 kg bomb and so damaged the ship's hull. The crippled vessel was towed to Leningrad under cover of darkness for repairs. Both Steen and his *Bordfunker, Uffz.* Alfred Schaarnovski, were killed. Other Soviet naval vessels sunk in these attacks were the flotilla-leader *Minsk* (2150 tons); the destroyer *Steregushchi* (1855 tons), which capsized after taking a direct hit; the submarines *M-74* and *P-2*, sunk in dock; the escort ship *Vihr;* the training ship *Svir;* and the patrol vessel *Taifun.* Among those damaged were the cruisers *Maxim Gorki* (8170 tons); the destroyers *Gordy, Grozyashchi* and *Silny* (1855 tons each); the submarines *Shch-302* and *Shch-306*; the depot ship *Smolny;* and the minelayer *Marti* (commanded by the former Prince Meshersky). (*Author's collection*)

Right: Well wrapped up against the bitter Russian weather, aircrew and ground crew getting this VII./St.G. 1 'Immelmann' *Stuka* ready for action are fighting both the Soviets and the Siberian wind chill on this bleak forward airfield in the winter of 1941/42. The unit's 'naval' emblem is still in evidence, as are the unit codings, but the *Balkenkreuz* and most other distinguishing features have been liberally daubed with white distemper to provide some suitable camouflage. The *Bordfunker* has dared to take his gloves off to adjust his machine-gun and pushed them in the rear handhold recess, but he should not be without them for long. The access panel to the port side engine coolant header tank is opened for heating – just getting the aircraft engines started was a major task at times. (*Ken Merrick*)

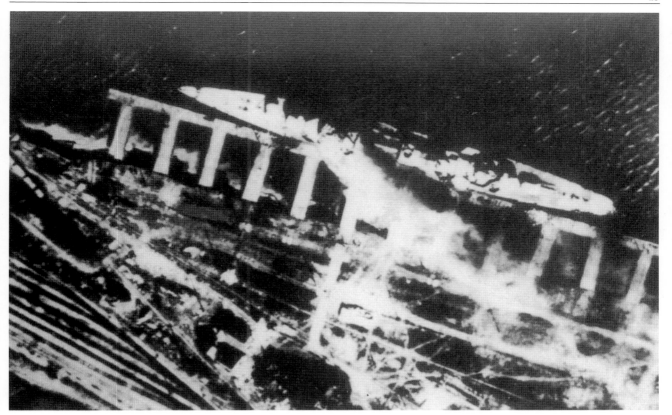

Above: On 18 March 1942 the aircrew of this Ju 87D have adopted suitable headgear to cope with the Russian weather as they prepare for a sortie. The pilot has left snowy footprints from his flying boots up the composite non-slip walkway on the wing root, while his rear-seat man manoeuvres his twin MG 81Z machine-gun back into place in his armoured turret after its overnight removal to stop it freezing up. The bitter winter weather had enabled the Soviets to go on the offensive and the *Stuka*s were flying missions to halt Russian spearheads from obliterating pockets of German troops. They held them but the damage had been done and German hopes now lay in the coming spring thaw and a new summer offensive. (*Archiv von Lutz*)

Opposite page, top: *Feldwebel* Pauser and *Bordfunker* Klose taking a break between sorties under the wing of their Junkers Ju 87B. The location is Kharkov-Rogan airfield, May 1942. The two SC-50 bombs on the under-wing racks are fitted with extended percussion detonator rods, which were designed to detonate the weapon above the ground and scatter shrapnel among the Soviet infantrymen. The bombs were called *Dinortstaben* after *Stuka* flyer Oskar Dinort, who was reputed to have introduced a makeshift homemade version of them during the Greek campaign, just prior to the Crete battle. Later they were produced in factories to a similar specification. (*Oelschlager via Sellhorn Archiv*)

Right: A stepped-up *Kette* of Ju 87B-2s of the VI./St.G. 1 over Russia in 1942, either training or returning from a mission with all bomb racks empty. The emblem is a red Tyrol eagle on a white shield – inherited from the original III./St.G. 165 at Wertheim via III./St.G. 51. Note the spinner marking of this unit. (*Archiv von Lutz*)

Left: Relaxed, pipe clenched in teeth, 'Blackmen' wheel the bomb-laden three-wheel loading trolley towards a Ju 87D, coded J9+AG, of VII./St.G. 1 on a forward airstrip in Russia, summer 1942. Note that the SC-250 bomb is fitted with four wind whistles. The unit emblem, a blue winged-helmet atop a red anchor on a white shield, is also featured on the driver's door of the Opal fuelling vehicle. This emblem was inherited from their days as the anticipated carrier-borne dive-bomber unit, I.(St.) *Trägergruppe* 186, originally destined for the first German aircraft carrier, the *Graf Zeppelin*, launched but never completed. The unit was re-designated III./St.G. 1 in July 1940. (*Archiv von Lutz*)

Opposite page, bottom: This Ju 87D-3 of the IX./St.G. 77, being 'bombed-up' on a main airfield in southern Russia, in the summer of 1942, wears its unit emblem unusually small. The 'Blackmen' are operating the hydraulic jack of the bomb trolley to cope with the large double fused SC-500 bomb in the foreground. The suspension eye-bolt can be seen on the metal band, as can one of the projecting fork-engaging lugs which slotted in the swing trapeze – designed to fling the main weapon well clear of the propeller arc during the near-vertical dive. (*Archiv von Lutz*)

Below: A 'Dora-3' of the VIII./St.G. 1 (*Werk Nr.* 2312), undergoes maintenance in Russia in 1942. The rudder appears to be painted in the *Staffel* colour of green. There is a gap in the rudder horn balance. The leather oleo covers have been removed from this machine but not her sister. The single tailplane bracing-strut, which replaced the double-strut of the 'Bertha', is shown to good advantage. (*Archiv von Lutz*)

Above: This collision between two 'Berthas' belonging to the 1. and the 3. *Staffel* respectively of the St. G. 77 took place at Sarabus-South airfield in the Crimea, during the summer offensive of 1942. One of the pilots, *Hauptmann* Scheffel, recalled later that the accident happened because the under-carriage of the 1. *Staffel* machine (on the left) had been shot away by AA fire while attacking Soviet 'Strongpoint 80' at Sevastopol. An emergency crash-landing followed, with the inevitable result, but there were no casualties from either aircraft, despite the fact that the 3. *Staffel* aircraft was fully 'bombed-up' and waiting to go. (*Immerreich via Sellhorn Archiv*)

Below: Junkers Ju 87Ds on patrol having recently joined the St.G. 2 in combat operations over Russia, in March 1942. There seems to be quite a varied mix of camouflage styles on these three aircraft. The Ju 87D closest to the camera car-rying the number '4' in white on her mottled wheel spat. This D-2 also differs from her two D-1 sisters in that she has been fitted with the metal glider-towing tail-box attachment below the rear fuselage and just under the rudder. This box could be either factory-fitted or added in the field to any *Stuka*, and was supported by a tubular metal hinged mount-ing frame which bolted on either side of the after lower fuse-lage. The towing cable was simply clipped on to this device. (*Archiv von Lutz*)

Above: Four Ju 87B-1s of the St.G. 2 'Immelmann' dispersed on a rough grass strip on the edge of the runway of a captured Soviet airfield in the Crimea in 1942. Note that none of them has the 'Trumpets of Jericho' sirens fitted to their wheel undercarriage at this time, nor is there any evidence of aircrew armour protection. (*Fritz Wenner via James V. Crow*)

Below: Men of the III.*Gruppe* St.G. 77 with their fighter escort leader discuss the next operation during the battles in the Crimea in the summer of 1942 (from the left): *Gruppenkommandeur Hauptmann* Helmuth Bode;

Jägerführer Oberst Molders; and Hans-Karl Sattler, the *Staffelkapitän* 8. *Staffel Oberleutnant. Hauptmann* Gerhard Bauhaus is in the foreground holding the map. Bode flew 300 combat missions in the *Stuka*, sinking a Soviet destroyer in the Black Sea as well as many other ships. He went on to become the *Staffelkapitän* of the experiment Ju 87 'Caesar' maritime *Stuka* during trials at Travemünde, and later became head of Ju 87 training in Bulgaria. Bauhaus carried out 482 *Stuka* missions before being fatally wounded during an attack near Rostov-on-Don on 22 July 1942. (*Frau Osten (Brand) via Sellhorn Archiv*)

Above: Many of the aerial photographs taken over Sevastopol, on which the *Stuka* relied so heavily for their target information, were taken by the short-range reconnaissance Focke Wulf Fw 189s, *Uhu* (Owl) aircraft, making the *Uhu* the forerunner of the U2. Here a 'Dora' keeps a close escort on one of these aircraft during such a mission in 1943. (*Niermann*)

Below: Over the rear tailplane of a *Stuka* as she pulls away from her attack dive on the enemy bunker positions south of Fort Stalin, the bursting bombs and resulting fire and smoke of the attack can be seen. These formidable defence works were being softened-up on 8 June, in readiness for the planned infantry assault the next day. (*Sellhorn Archiv*)

Opposite page, top: The attack on Sevastopol harbour was resumed on 9 June 1942 by the *Stuka*s of St.G. 77. Under the wing-tip of this I./St.G. 77 *Stuka* can be seen the town itself with its inlets and beyond, in the distance, the main Soviet naval base. (*Sellhorn Archiv*)

Opposite page, bottom: Crimea, the summer of 1942. A *Stuka* seen turning away after a dive-bombing attack, north of Severnaya Bay, on 9 June 1942, leaving fire and smoke in its wake. The wheel spats of the undercarriage have been removed from this Ju 87 of St.G. 77. (*Sellhorn Archiv*)

Opposite page, top: The Soviet floating AA battery, PZB-3, *Ne Tron Meniya*. This strange vessel was a floating experimental section taken from the material assembled at Nikolayev on the Black Sea for a new 59,150 ton battleship, to be named *Sovetskaya Rossiya*. After the German attack the decision was taken to convert this section into a floating battery to defend Sevastopol airfield from attack on its seaward side. The battery was armed with three 103mm, three 76mm, three 37mm flak guns and four machine-guns. On 8 August 1941 it was towed into position off the lighthouse at Cape Chersonese. Commanded by Captain-Lieutenant Sergei Moshenskii it survived numerous air attacks despite frequent damage. In January 1942 it broke away from its moorings and drifted towards Balaclava, being taken in tow in March of that year, at Kazach Bay. Following further damaging *Stuka* attacks, of which this, conducted by St.G. 77, was one, this floating fortress was again abandoned on 27 June 1942. (*Sellhorn Archiv*)

Above: Deep discussion of a technical problem absorbs the interest of these officers of the *Stabstaffel* St.G. 77 during the summer of 1942. The unit's motor transport in the foreground, the Kfz 2/40, was the *Luftwaffe*'s equivalent to the Allied Jeep, a multi-purpose vehicle used to move around rough terrain. The broad arrow pointing downward atop the aircraft's nose, just in front of the pilot's cockpit, indicates the oil tank cover. The circular cover just in front of the red warning triangle marks the oil filling point. (*Dr. Armbrust via Sellhorn Archiv*)

Left: Coming in to land at an airfield in Russia in 1942 these two 'Dora-1s' of the VII./St.G. 2 bank sharply to the left and show their underwing markings. The aircraft closest to the camera carries T6+HR on her starboard fuselage, with the 'R' in yellow. Note the distinctive markings on the top of the rudder of the far-side machine, which probably indicate that the aircraft belongs to the *Staffelkapitän*. (*Archiv von Lutz*)

Opposite page, left: The ceremony to mark the award of the *Ritterkreuz* to the *Staffelkapitän* of the 2. *Staffel* St.G. 77, *Hauptmann* Helmut Leicht. The advance on Stalingrad and the Volga river was in full swing when this photograph was taken on 3 September, and once more the *Stuka*s are in full cry and spirits are high. Among those present are *Feldwebel* Herbert Rosenhauer (third from left); *Uffz.* Anton Pronold (fifth from left); *Flugzeugführer Feldwebel* Sepp Haber (sixth from left); *Staffelkapitän Hauptmann* Helmut Leicht himself (with shield and garland); and *Uffz.* Beno Fischer (fourth from right). (*Graf via Sellhorn Archiv*)

Left: A photograph of Ju 87D-3s of the IV./St.G. 77 returning from a sortie over the Nikolayev area of southern Russia in 1942. The aircraft nearest to the camera is coded S2+NM. This photograph gives a good view of the round plexiglass access cover on top of the main body of the aircraft's fuselage abaft the rear cockpit. This shielded the *Peilfunkgerät* direction-finding equipment held in a recess below. (*Archiv von Lutz*)

Above: 'Doras' of the II./St.G. 2 in Russia during the summer of 1942. These aircraft still display their unit markings and embellishments, from the pilot's girlfriend's name 'Bärli', on the nose, to the unit emblem showing the *Bamberger Reiter* (Knight of Bamberg) and the *Werk Nr.* 2461 atop the white highlighted swastika on the tail. Note that 'Bärli' carries a 'Trumpet of Jericho', complete with propeller, on her port undercarriage leg only. These wind-powered sirens re-enforced the screaming sound of a dive-bomber attack, and were used to inspire terror in troops on the ground. (*Archiv von Lutz*)

Left: *Stuka*s over Stalingrad. This *Kette* of dive-bombers from St.G. 2 banks over the snow-mantled city with the Volga river beyond. Although some claim the *Stuka*s did not fly 'close-support' missions, the truth is that nothing flew closer air support than the Ju 87. Hans-Ulrich Rudel, whose unit was in the thick of the fighting, recorded how, 'Each pilot is given his target precisely marked with a red arrow. We fly in, map in hand, and it is forbidden to release a bomb before we have made sure of the target and the exact position of our own troops.' (*Signal via author's collection*)

Opposite page, bottom: Dawn on a frosty Arctic airfield in northern Finland during the winter of 1942/43. A 'Dora' taxies out ready for take-off watched by a 'Blackman'. In the foreground, snow-covered bombs lie ready, while another aircraft of the same group has a single underwing SC-250 bomb on the faired underwing side rack. (*Richard Chapman*)

Above: A bleak outlook on this forward base during the winter of 1942/43. These long-range 'Richards' are parked out in the open with little camouflage from the wintry conditions; the *Stuka* on the left is running up her engines to keep them ticking over in the sub-zero temperatures, which could hit as low as forty degrees below zero. The portable heating equipment had a large diameter flexi-hose which connected to the

curved fitment on the top and blew warm air into the engine compartments. Special anti-freeze mingles with bomb crates and bombs ready for loading for the next mission. (*Ullstein*)

Below: *Leutnant* Gerhard Baunacher's long-winged Ju 87D-5 on a snow covered runway in Russia in the winter of 1943/44. The aircraft carries her individual aircraft letter 'B' on the front of both wheel spats. This photograph gives a good view of the coolant radiators located on the inboard sections under both wings of the 'Dora'. Here the wings have their rectangular access flaps fully lowered. Behind the *Stuka* the awesome bulk of a mighty six-engined Me 323 transport comes down to make a landing. (*Gerhard Baunacher via James V. Crow*)

Opposite page, top: Men of the St.G. 77 outside their billets at Taman, which they had only just recently occupied, January 1943. In the front row, from the left, can be seen *Flugzeugführer Ofzw.* Alois Wosnitza, *Staffelführer Oberleutnant* Theodor Haker, *Bordfunker* Milo Hauf, *Flugzeugführer* Ernest Kelle, an unknown *Bordfunker* and *Bordfunker Uffz.* Maahs. Behind them, from the left, are *Flugzeugführer* Hans Zellner, *Flugzeugführer* Siegfried Huber, and *Bordfunker* Gunther. Three of these pilots were great *Stuka* pilots and were to be awarded the Knight's Cross: Haker (20 February 1944), Huber (3 April 1943) and Wosnitza (26 March 1944). (*Maahs via Sellhorn Archiv*)

Opposite page, bottom: A Ju 87D-5 of the VII./St.G. 2 on a forward airfield in the Crimea, late spring 1943. This is the aircraft of the *Staffelkapitän, Leutnant* Hubert Pölz (on the left). Note that both wheel spats have been removed

evidently because of the churned-up condition of this airstrip. Pölz carried out a total of 1055 combat missions. Of these 704 missions were in the *Stuka*, before moving on to the Fw 190. He was shot down four times, wounded three times and destroyed eleven Soviet fighters, as well as sinking the Australian sloop *Auckland* off Tobruk harbour and three Soviet destroyers in the Black Sea. (*Helmut Mohr via James V. Crow*)

Above: *Major* Paul-Werner Hozzel at the controls of his Ju 87D-3 of the *Stab./St.G.* 2 in Russia in 1943. His machine features the *Kommandeur*'s chevron in front of the unit emblem as well as the distinctive white spiral on the propeller boss. The exhausts have the long, open-ended, frontal shroud. This aircraft was flown carrying both undercarriage wind sirens, with their associated propellers, in place. (*James V. Crow*)

Opposite page, top: *Stab* I./St.G. 77 flies over Kharkov in 1943. The aircraft is unarmed and only carries one auxiliary fuel tank under the starboard wing. The *Stuka*s of this unit were heavily involved in the various battles for control of this city, which changed hands frequently that year. They operated for some time from the north and the south airfields, supporting the army both in defensive fighting, to hold the Soviet breakthrough early in 1942, and the offensive towards Kursk, which followed. (*Sellhorn Archiv*)

Opposite page, bottom: A Junkers Ju 87D-3 of the VII./St.G. 1 casts her shadow over the steppe below as she makes a low run back to base at the end of another mission. This early model still retains the fairing for the wind-driven siren but these were rarely used by this date. Her port-side ETC 50/VII underwing racks can be seen and she carries the identifying letter 'N', repeated on her starboard wheel spat. (*Archiv von Lutz*)

Above: Despite its dappled camouflage this Ju 87 'Bertha-2' has been caught on the ground and well strafed and eliminated. Its midships fuselage is now a burnt-out tangle and its starboard tailplane is full of cannon shell holes. Note the twin horizontal tail 'vee-braces' and that the cover to the starboard Rheinmetall-Borsig MG 17 wing gun bay access has been removed. (*Harold Thiele*)

Below: A Ju 87G-1 VzP *Kanonenvogel* ('cannon bird'), another term for this variant of the *Stuka*. This aircraft belongs to the 10.Pz./SG. 3. Converted from the D-3, these early versions of what was to become the *Luftwaffe*'s main tank-destroying aircraft, World War II's equivalent of the Fairchild Republic A-10 Thunderbolt-II. The redundant dive-brake struts can be seen under the leading edge of the wing. (*Archiv von Lutz*)

Left: Fatal accidents add to the normal toll of war. This *Stuka* of 1.*Staffel*, St.G. 77 was forced to make an emergency landing near Kharkov due to the death of a radioman from a 'blow-back' explosion in an MG 81Z machine-gun. (*Dr. Armbrust via Sellhorn Archiv*)

Below: A potential replacement for the Junkers Ju 87 *Stuka* which didn't quite make it. This is a Henschel Hs 129B, a twin-engined ground-attack aircraft, undergoing combat testing at the front in 1943. Based on the Soviet Il-2 Sturmovik attack plane, with almost the entire centre section of the aircraft and the cockpit 'box' made of welded armour plate, it was fitted with French Gnôme-Rhone 14M radial engines, and armed with two MG 151/20 cannon and two 7.9mm MG 17 machine-guns as well as a variety of under-wing weapons. Later models were fitted with a single 30mm MK 101 cannon with thirty rounds of armour-piercing tungsten tipped shell for tank busting. The engines proved unreliable and not rugged enough to face the variations of the Russian weather, making for poor serviceability. Yet again, although in theory superior to the *Stuka*, this aircraft supplemented, but never replaced, the Ju 87 in service. (*Dr. Armbrust via Sellhorn Archive*)

Right: The main base of the *Gruppenstab* of the I./St.G. 77 at Kharkov. A Ju 87D of that unit (on the right) can be seen in a 'splinter-box' next to a Junkers Ju 52 transport aircraft in the next bay. The Ju 52 has just left the compass-swinging and alignment position set in the ground at the bottom centre of the photograph. (*Dr. Armbrust via Sellhorn Archiv*)

Below: A chevron adorns the front of the unit emblem of the black 'Scottie' on a red disc on a 'Dora' belonging to *Leutnant* Egbert Jaekel, *Staffelkapitän* of the II./St.G.2 'Immelmann'. Those who believe that the *Stuka* stood little chance against Allied fighters should note that Jaekel was credited with knocking down ten or eleven Soviet fighter aircraft in his *Stuka* during his combat career. He lost his life when, on 17 July 1943, he got into a dog-fight with a forma-tion of Lavochkin La 5s. Having disposed of his twelfth fighter, Egbert was recovering at low level (200 meters) when he was jumped by several more. Heavily hit and without the height to recover, his aircraft went straight down, burning on impact and killing him and his rear-seat man, *Oberfeld-webel* Fritz Jentzsch, who is seen here joking with his mechanics shortly before his final mission. (*Hans Obert*)

Above: The pilot of this 'Dora' strains to hear the shout of his ground mechanic above the roar of the Jumo 211 J engine on a makeshift airstrip close to an orthodox church in the background. This photograph gives a good view of the new supercharger air intake, with its cover open flat against the nose of the aircraft, and the offset two-third/one-third interior division of the new streamlined oil cooler intake. The lever-activated vents were closed during the dive. (*Archiv von Lutz*)

Below: The lean profile of a Junkers Ju 87D-1 of the III./St.G. 1 as it banks over the delta of the Dnieper river during the bitter autumn battles of 1943. With the loss of Kharkov and its adjacent airfields on 23 August 1943, the *Luftwaffe* withdrew ground-support units to airfields further south in order to hold the line of the Dnieper and protect the flank of the Crimea. This meant that III./St.G. 1 found itself flying from Kirovograd and Krivoy Rog. (*Ken Merrick*)

Above: The Hungarians bought twelve D-1 and D-3s in 1943, and, a year later, a final batch of fourteen D-5s. Both D-3s and D-5s were used to equip their solitary dive-bomber unit, 102/1 *Zuhanobombezo Szezad*, which had a complement of twelve machines in 1944. This war-weary Ju 87D-5, with the individual code number 6, is pictured in Hungary early in 1945. The fuselage band and the wing-tips are white. The complete wheel spatting has been removed to facilitate landings and take-offs in the muddy conditions then prevalent. (*Denés Bernád Archiv*)

Right: *Hauptmann* Heinrich Zwipf on the non-slip metal bands at the wing root of his *Stuka*. This photograph was taken at Kharkov airfield during the period of German defensive operations in 1943. Note the sliding canopy, partly-armoured pilot's seat, with the rotating plate open, and the aircrew hand-grips. Zwipf served in the Polish campaign in 1939 before becoming a *Stukaschule* instructor. He was the *Staffel-kapitän* of the III./St.G. 77 from September 1942 until 19 October 1943 and flew over 600 combat sorties in the *Stuka*. He was later killed flying Fw 190s with the I./SG. 4 in Italy in April 1944. (*Sellhorn Archiv*)

Opposite page, top: The veteran team of (centre and right) *Hauptmann* Heinrich Zwipf, *Staffelkapitän* III./St.G. 77, and *Feldwebel* Heinz Sellhorn. The occasion is Zwipf's 500th combat flight celebration in the summer of 1943. The pair went on to complete more than 600 missions each by October of that same year. Note the outline 'N' stencilled against the fuselage band carried by their mount at this period. The woven matting fences had to serve as bomb-splinter revetments for the *Stuka*s in this part of the steppe. (*Sellhorn Archiv*)

Opposite page, bottom: New weapons to fight the powerful Soviet T-34s. *Hauptmann* Heinrich Zwipf, the *Staffelkapitän* of III./St. G. 77 is here seen watching a maintenance inspection by one of his ground crew for the newly fitted Mauser MG 151 wing cannon. Two of these 20mm (or 2cm) long-barrelled weapons were carried to further aid the *Stuka*'s ground-strafing role. They replaced the stubby MG 17 machine-guns carried in the same positions. The new weapons were sometimes fitted with optional flash muzzles on later Ju 87D-5s. (*Sellhorn Archiv*)

Above: A perfect three-point landing by the *Gruppenkommandeur* of I./St.G. 77 *Hauptmann* Karl Henze, in his 'Dora' back at Kharkov airfield in 1943. Used to operating in the mud or dust of the Russian steppe, the complete wheel spatting has been removed from his aircraft. The luxury of a proper runway on which to touch down was a rarity. (*Dr. Armbrust via Sellhorn Archiv*)

Below: This long-winged dapple-camouflaged 'Dora-5' is the object of considerable interest to the assembled Russians on this airstrip and may be one of the first they have seen. The *Werk Nr.* is carried below the swastika but unfortunately is illegible in this photograph. The oleo covers have been removed. This variant formed the mainstay of the redesignated *Schlachtgeschwader* ('ground-attack') units following a general review in October 1943. (*Archiv von Lutz*)

Left: The *Kommandeur* of the II./St.G. 77, *Major* Helmut Leicht (left) receiving a garland on the occasion of his 600th combat sortie, in October 1943. He was to survive many more *Stuka* missions only to go missing in June 1944, while flying a 'less vulnerable' Fw 190 fighter-bomber, in sorties against Soviet tank columns at Vitebsk. The *Unteroffizier* in the centre is wearing the standard informal service dress of front-line personnel, a blue 'fighter blouse' with silver epaulettes and collar, side-cap with the *Luftwaffe* eagle, service dress trousers and black boots. (*Sammler Kroll via Sellhorn Archiv*)

Below: Winter warpaint for these Ju 87D-1s of St.G. 1. (Background) Whitewash has been dappled over the fuselage, undercarriage and other upper surfaces, obliterating unit emblems. However, the J9 coding and the *Balkenkreuz* remain at full-size as does the swastika on the tail. The swastika appears within a square of the original paint scheme. (Foreground) The 'Trumpets of Jericho' still appear on both undercarriage legs even as late as the winter of 1942/43. Clearly visible is the redesigned swing trapeze and adjustable strap to enable the 'Dora' to carry an increased main payload, although the SC-250 and SC-500 remained the principal ordnance. (*Harold Thiele*)

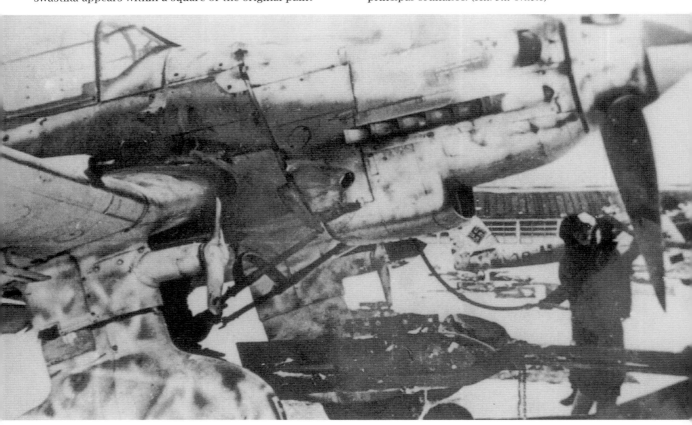

Right: Under the stern gaze of the Führer the 'top circle' of SG. 77 relax between operations at Kalinovka, Russia, in November 1943. At the back (from the left) are *Leutnant* Mueller, the unit's signals officer; an unknown officer; and the *Gruppe Adjutant*. At the front (from the left) are the Ic of SG. 77, *Hauptmann* Gohmann; the *Gruppen-kommandeur* I. SG. 77, *Major* Karl Henze; *Hauptmann* Luneburg, *Staffelkapitän* of II./SG. 77; *Hauptmann* Hans-Joachim 'Cherry' Brand, *Staffelkapitän* I./SG. 77; Dr. Egerndorfer, *Gruppe* medical officer; and an unknown *Leutnant*. (*Frau Osten (Brand) via Sellhorn Archiv*)

Below: A Ju 87D-5 of the I./SG. 3, fitted with a glider towing attachment, and carrying a full load of bombs, follows a rail line through the snow-bound landscape. The unit had left the Mediterranean after winning the Aegean campaign in the autumn of 1943 when they again inflicted heavy losses on the Royal Navy and drove the British from Leros and Cos in a replay of the battle for Crete two years earlier. Now the unit operated on the northern sector of the Russian front, commanded by *Hauptmann* Hans Topfer. They were among the *Schlachtfliegergeschwader* desperately trying to stem the Soviet tidal wave sweeping towards Germany. (*Archiv von Lutz*)

Above: *Generalleutnant* Hans Seidemann (centre) inspects the I.*Gruppe* SG. 77, with the new *Kommandeur*, *Hauptmann* Hans-Joachim Brand (left) in November 1943. Behind them is *Geschwaderkommodore Oberstleutnant* Helmut Bruck. Seidemann had been Richthofen's chief of staff in both the *Legion Condor* and VIII.*Fliegerkorps* and was an expert at close air support. Seidemann gave the Allies a bloody nose in Tunisia with his handful of *Stuka*s, before returning to the Eastern Front, this time as commander of VIII.*Fliegerkorps*, which he led to the end of the war. (*Sirrenberg via Sellhorn Archiv*)

Left: A good in-flight photograph of the Ju 87Ds of III./St.G. 77 operating in southern Russia in 1943/44. All badges and insignia have vanished under the liberal coating of whitewash, and non-stop operations allowed little time for the spick-and-span attention that had characterised the earlier years of the war. From their selfless devotion to helping their hard-pressed comrades on the ground, the 'flying artillery' was dubbed 'flying infantry'. (*Otto Schmidt, via Sellhorn Archiv*)

Right: SG. 77 *Gruppen-kommandeur Major* Karl Henz (centre) of the I./St.G. 77 on the occasion of his 1000th combat sortie on 26 February 1944, at Uman airfield in Russia. *Hauptmann* Wolfgang Schenk can be seen on the far right. He went on to record 1098 combat missions and survived the war as *Kommodore* of an auxiliary *Geschwader*. (*Sirrenberg via Sellhorn Archiv*)

Below: On achieving his 1000th *Stuka* combat mission on 7 March 1944, when he flew due north from Orsha airfield to destroy Soviet artillery south-east of Vitebsk, Byelorussia, *Major* Friedrich Lang, the *Gruppenkommandeur* of III./SG. 1, is met in traditional manner, and presented with a 'Jolanthe' (the *Stuka* mascot) by an appropriately garbed 'Blackman' in a stovepipe hat. Such incredible combat sorties were quite normal for *Luftwaffe* personnel and *Stuka* men in particular. There were no tours of duty for them, unlike their Allied counterparts. The *Luftwaffe* did not have limitless manpower, and had to fly continually to help their comrades on the ground. (*Ken Merrick*)

Above: Illustrating the basic types of *Luftwaffe* clothing worn by *Stuka* men during the war are (from left to right) *Ofw.* Werner Honsberg, *Oberleutnant* Anton Andorfer and *Uffz.* Toni Kellner, all of St.G. 77. Andorfer is wearing the standard service dress uniform, with his tunic collar and the seams of his peaked cap piped silver, with silver cords, to show his officer ranking. The epaulettes were silver on yellow for aircrew, while collar patches were piped silver with yellow wings and oak spray denoting an *Oberleutnant* rank. On his left breast, above the pocket, is a gold mission clasp. On the pocket he wears an Iron Cross, 1st Class, awarded to him on 26 March 1944, after he had completed 726 combat missions. Below it is his pilot's silver qualification badge. On his right breast is the silver *Luftwaffe* eagle, and from the centre button hole the ribbon of the Iron Cross 2nd Class is displayed. The two officers on either side of him are wearing the standard light brown canvas flying suit zipped diagonally. Hondsberg has a rank patch with four wings on his left arm and has a standard lightweight canvas flying helmet with earphones made of bakelite. Kellner is wearing the side-cap. This photo was taken at the Nikolayev airfield, on the Black Sea estuary of the Bug river, in June 1943. (*Hondsberg via Sellhorn Archiv*)

Left: A close look at the Junkers Ju 87D 'Dora' on 14 April 1944. The supercharger air intake was fitted with an opening and closing flap at the front (here seen fully open), which shut for the attack dive. The reduced, shallower oil cooler intake replaced the deep radiator of the 'Bertha' and the two radiators were replaced under each wing, inboard. As can be seen here the new, stronger undercarriage also carried a much reduced profile with the wheel spats made smaller, thinner leggings and the oleo compression bands under thin leather outer coverings. (*Göhmann via Sellhorn Archiv*)

Above: Here the Ic of SG. 77, *Hauptmann* Göhmann, congratulates *Hauptmann* Helmut Bruck, the *Geschwaderkommodore,* on the occasion of his 900th combat sortie, at Lemberg airfield (now Lvov in Poland). Bruck had served with the *Stuka*s since joining the I./St.G. 165 at Kitzingen in April 1936, and had fought in Poland (surviving being shot down on the first day), the Netherlands, Belgium, France, and in the fighting over the Channel. After becoming *Kommandeur* of I./St.G. 77 he fought in Yugoslavia, Crete and then on the Russian front, making 958 *Stuka* combat missions and ending the war as *General der Schlachtflieger Nord.* (*Göhmann via Sellhorn Archiv*)

Opposite page, top: The 600th combat mission of *Feldwebel* Horst Goertler (in the centre) of *Geschwader* SG. 77 was celebrated on 16 April 1944 at Lemberg air base. Sharing his celebration are (from the left) *Bordfunker* Willi Diessel; *Flugzeugführer Uffz.* Berzlmaier; *Waffenmeister Feldwebel* Woerle; and two unknown *Unteroffiziers*. Goertler served with both the St.G. 2 and the St.G. 77 with distinction and fought with the SG. 77 until the war's end, surrendering to the Americans in 1945. They gave this brilliant pilot over to the Soviets, who sent him to Siberia. Here he developed pneumonia and almost died. He returned home later and indomitably resumed flying, dying when his sports plane spun-in at Lorsch/Hesse on 20 October 1956. (*Berzlmaier via Sellhorn Archiv*)

Opposite page, bottom: April 1944 and the men of *Stukageschwader* 77 celebrate an incredible 100,000 combat missions since the start of the war, as the plaque held on the spinner of the 'Dora' proudly shows. On the far left is *Hauptmann* Heinz Niehus, *Staffelkapitän* of VIII./SG. 77, and later appointed acting *Kommandeur* of III./SG. 77. Next to him is *Oberleutnant* Heidle and they are surrounded by various unknown *Feldwebels*. (*Göhmann via Sellhorn Archiv*)

Below: A scene at Lemberg airfield on 14 April 1944. The Soviets are pressing hard and the *Stukas* are still in constant action trying to stem the flood. 'Blackmen' fuel up the aircraft of SG. 77, working from the mobile tankers at this home base that was now close to the front line. The unit's *Kommodore, Oberstleutnant* Helmut Bruck, was to fly his 900th sortie from this field on 5 June 1944, just one day before the Western Allies landed in Normandy and opened the second front. (*Göhmann via Sellhorn Archiv*)

Below: On the Romanian airfield of Husi, *Leutnant* Hans Koslowski (on the left) watches as *Hauptmann* Hendrik Stahl, the *Staffelkapitän*, briefs the aircrew of the 8. *Staffel* of III./SG. 2 for another mission against Soviet tank columns thrusting hard for the border. Stahl, along with his radio operator, *Oberfeldwebel* Georg Poethig, flew more than 1200 combat missions in the *Stuka* from the first day of the Russian campaign to the very last day, and was credited with the destruction of 66 enemy tanks as well as three Soviet fighters, an ammunition train and twenty-five flak batteries knocked out. This was in addition to sinking one British destroyer and damaging another in the Balkans campaign. (*Hans Koslowski via James V. Crow*)

Opposite page, top: A Ju 87D-3 seen over the Eastern Front in 1944. Both underwing bombs have the long 'Dinortstab' extensions fitted. The aircraft is fitted with a glider towing attachment under its rear fuselage. By now sirens had been eliminated from the production line. Only 1559 D-3s had been built between May 1942 and June 1943. The gliders that they towed were mainly full of supplies and vital parts rather than being troop carriers. (*Archiv von Lutz*)

Opposite page, bottom: The long-barrels of the Mauser MG 151 20mm cannon show up here as this *Feldwebel* poses on the wing of a 'Dora-5' in Poland during the summer of 1944. Each cannon had 500 rounds and these were carried in a curved metal container situated flat within the wing outboard of the gun's firing mechanism. The upper wing surface access panel to re-arm and service these weapons can be seen opened on the extreme left. The aircraft carries her identification codes – the letter 'B' on the front, and number '2' on the side, of her wing spats. (*Ken Merrick*)

Below: A good range of heavy bombs awaits selection by the armourers of the 8. *Staffel* III./SG. 2 on the Romanian airfield of Husi in 1944. Note that the bomb closest to the aircraft has wind-whistles attached to the fins. The nearest *Stuka*, coded T6+LS with the 'L' in red on the yellow fuselage band, is the aircraft of *Leutnant* Hans Koslowski and his *Bordfunker*, *Uffz.* Horst Klein. (*Hans Koslowski via James V. Crow*)

Right: Kostrompka airfield in the Ukraine on 31 May 1944. The 500th combat mission of *Feldwebel* Heinz Meyers (centre, holding the flowers) of the 8. *Staffel* III./SG. 2. Watched by a crowd of well-wishers in front of his *Stuka*, he is congratulated by *Leutnant* Hans Koslowski (in short sleeves to the left) and *Oberleutnant* Hendrik Stahl, the *Staffelkapitän* with *Ritterkreuz* (to his right). Meyer was awarded the German Cross in gold as *Fahnenjunker-Feldwebel*. He ended the war having flown a total of 618 combat missions, destroying 40 Soviet tanks, 50 artillery pieces and more than 100 trucks and motor vehicles. (*Hans Koslowski via James V. Crow*)

Above: The award of the *Ritterkreuz* to *Ofw.* Werner Honsberg (second from the left) of the I./St.G. 77 on 20 July 1944, at Seisdorf in Silesia. Pictured here are *Major* Karl Henze (on the left) of I./St.G. 77, *Ofw.* Werner Honsberg and *Hauptmann* Hans-Joachim Brand (on the right), *Staffelkapitän* I./St.G. 77. (*Sellhorn Archiv*)

Left: A huddled pre-mission briefing is photographed on 27 April 1944, on a forward airstrip, while a bombed-up Ju 87D-5 waits to take off. This gives an excellent view of the tapered wing-tip extensions which were a feature of this model. They increased the *Stuka*'s wingspan to 49 feet 2 inches overall, which improved the aircraft's wing loading. The wing guns are long-barrelled Mauser MG 151 20mm cannon. The belted ammunition was usually armed with alternate high-explosive and armour-piercing rounds. (*Archiv von Lutz*)

Right: A Ju 87G-2 of the 10.Pz./SG. 3 on a forward airfield in Latvia, August 1944. Seen talking to the pilot, *Uffz.* Klemme, is *Luftwaffe* war reporter *Leutnant* Speck. In March 1944 the 4./St.G. 2 converted into the 10.Pz./SG. 3 and from 7 January 1945 operated as the *Panzerjägerstaffel* of the SG. 3 under the command of *Oberleutnant* Andreas Kuffner. Based on the Baltic coast, it defended East Prussia and gave time for the evacuation of German civilians from the path of oncoming Soviet armies. (*Konrad Honika via James V. Crow*)

Above: Officers of the SG. 77 and their families enjoy an outdoor meal at Seifersdorf airfield, Silesia, in August 1944. Those present include (from the left) Frau Sagan; *Zahlmeister* Sagan; Frau Luneburg; *Staffelkapitän* 1. *Staffel* SG. 77 *Hauptmann* Brand; Frau Bruck; *Geschwaderkommodore* SG. 77 *Oberst* Helmut Bruck; *Hauptmann* Luneburg with his daughter; and Frau Brand with her daughter. (*Sellhorn Archiv*)

Opposite page, top: The rear-seat men became redundant when the *Stuka*s finally began to be replaced in most units by the single-seater Fw 190s from June 1944 onwards. Radio operators and rear-gunners were therefore returned to re-allocation centres in Germany for fresh assignments. A few became pilots in their own right, many transferring to *Nachtschlacht* units, while others were placed in ground units. A group of SG. 77's *Bordfunker*s, their *Stuka* flying days over, are pictured here at one such centre, Lubben, south of Berlin, in the autumn of 1944. (*Sellhorn Archiv*)

Right: It is one of the ironies of the war that many *Stuka*s of the *Fortele Aeriene Regale ale Romaneiei* (Royal Romanian Air Force) were turned against their former allies when Romania switched sides. From August 1944 German troops also had to duck when the inverted-gull wing shape appeared in the sky. The first such attack took place near Cluj-Napoca in Transylvania. A Romanian Junkers Ju 87D-5 equipped with anti-personnel bombs is prepared for a mission in the summer of 1944. (*Denés Bernád Archiv*)

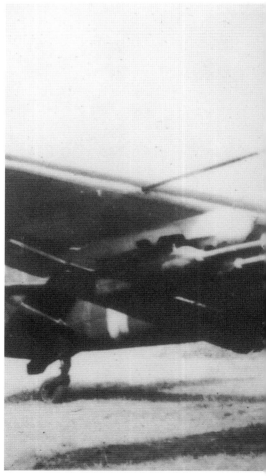

Opposite page, top: The wreckage of the Ju 87G-2 (*Werk Nr.* 494231) of *Feldwebel* Joseph Blümel of the 10.Pz./SG. 3. Blümel was shot down by Soviet flak machine-guns at 11.10 hours on 19 September 1944, eighteen kilometres to the south of Riga, while on his second mission of the day. He had just destroyed his 60th enemy tank on the first mission but the damaging hit forced Blümel to make a crash landing behind enemy lines at Kekava. Both Blümel and his *Bordfunker*, *Obergefreiter* Hermann Schwarzel, survived the crash but were shot by Soviet troops. Their bodies were found three days later by a search party led by the *Staffelkapitän, Hauptmann* Andreas Kuffner, and buried with full military honours. Blümel was awarded the Knight's Cross, posthumously, on 28 January 1945. (*Wilhelm Landau via James V. Crow*)

Left: 'Blackmen' in a procession with the main bomb on the hydraulic arm of the loading trolley approach the 'Dora-5' aircraft of *Leutnant* Gerhard Baunacher of the I./SG. 1 on a snowy airstrip in Russia, while he goes through his cockpit check routine. The flexible adjustable arms of the bomb cradle hang down in readiness for their load, while yet more armourers are working on the underwing racks. (*Gerhard Baunacher via James V. Crow*)

Above: *Feldwebel* Qualo of the 10.Pz./SG. 3 prepares for another mission at his forward base in East Prussia, late in 1944. He is loading his twin flexible 7.92mm MG 81Z machine-guns back into his GSL K-81 gun turret after they had been unshipped for servicing. This photograph gives an excellent view of this weapon. Note that the wing-root walkway now consists of three riveted metal strips instead of the earlier composite non-skid surface. (*Willi Bromen via James V. Crow*)

Opposite page, top: *Leutnant* Gerhard Baunacher of the I./SG. 1 with his 'Dora-5' standing ready on a snow-bound airstrip while the armourers struggle with the fin adjusters of her main payload. She carries the personal aircraft identification letter 'B' on the front of both wheel spats. After several years of experience of winter combat conditions on the Eastern Front all 'Doras' were equipped with jettisonable landing gear. (*Gerhard Baunacher via James V. Crow*)

Left: The hand-over of the command of SG. 77 to the new *Kommodore*, *Oberstleutnant* Manfred Mossinger, at Krakow, Poland, on 18 November 1944. Those present are (from left to right): *Geschwaderkommodore* SG. 77 *Oberst* Helmut Bruck; Ic SG. 77 *Hauptmann* Göhmann; and *Gruppenkommandeur* II./SG. 77, *Hauptmann* Alexander Glaeser. (*Dittewig via Sellhorn Archiv*)

Above: The appointment of *Oberstleutnant* Mossinger of SG. 77 is celebrated by fellow officers at Krakow airfield on 18 November 1944. Seen from the left are: *Oberst* Helmut Bruck, *Kommandeur* SG. 77; *Oberleutnant* Dr. jur. Maximilian Otte; *Gruppenarzt* I./SG. 77 Dr. Egerndorfer; *Gruppenkommandeur* I./SG. 77 *Hauptmann* Hans-Joachim Brand; and *Staffelkapitän* III./SG. 77, *Oberleutnant* von Kathen. (*Frau Osten (Brand) via Sellhorn Archiv*)

Above: A 'Gustav-2' of the 10.Pz./SG. 3 seen at Ulez airfield, Poland, in the winter of 1944/45. Perhaps the most striking feature of this photograph is the extreme youthfulness of the aircrew. Even at this late stage of the war the individual aircraft coding letter 'E' is being carried on the front of the wheel spats. (*Konrad Honika via James V. Crow*)

Opposite page, top: Line-up of three 'Gustav-1' *Panzerknacker*s on a snow-covered airfield of Ulez, Poland, in the winter of 1944/45. The individual aircraft identity numbers '4', '5' and '2' are shown in stark white against their dark wheel spats but no camouflage appears to have been attempted to deal with current conditions, despite the fact the Germans now had their backs very much against the wall. (*Konrad Honika via James V. Crow*)

Right: Two 'Blackmen' cranking up the starter motor of this Ju 87G-1 'Gustav', of the 10.Pz./SG. 2, *Staffelkapitän Leutnant* Anton Korol's aircraft. This ace tank-buster led the unit from 1 September 1944, until the end of hostilities. He killed 98 Soviet T-34s and one Stalin tank, and damaged 200 others, for which he was awarded the Knight's Cross on 12 March 1945. His aircraft carries the code 'B2' on each wheel spat. The G-1 variants were converted from existing late-model D-3s and could be identified by the retention of the MG 17 machine-gun fairings on the wings. Some still had the superfluous dive-brake bracing in place beneath each wing. (*Archiv von Lutz*)

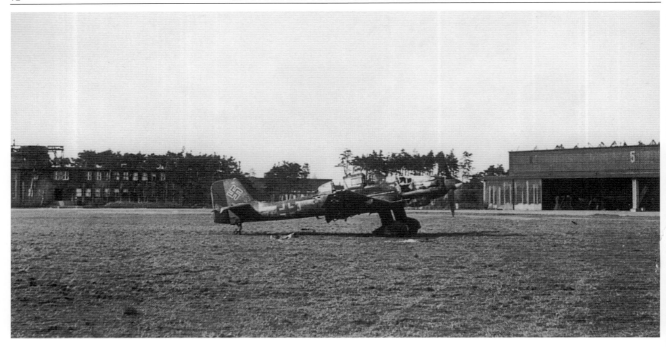

Above: The end of the war and this damaged and stripped-down Ju 87D-5, one of 59 located by the British Disarmament Wing, stands abandoned on a German airfield now occupied by the Allies. The *Stuka* imparted such a legacy of dread that each one was carefully and deliberately destroyed to a precise and detailed plan. Only two survived intact from a grand total of 5709 aircraft. The much-feared, much-derided but multi-talented *Stuka* had fought from the first to the last day of the war, despite attempts to write it off by the Allies, or pension it off by the Germans. As flying artillery, helping to win battles in weeks rather than years, and its pioneering roles as an aerial tank- or ship-destroyer,

the slow but accurate – and later much imitated – Ju 87 deserves its place as one of the most famous warplanes in history. (*Imperial War Museum*)

Below: Having fought Soviet tanks to the very last day, a number of *Stuka* units chose to fly their aircraft and surrender them to the British or Americans. This 'Gustav' is one of two that flew in from Czechoslovakia and landed at Eschwege airfield, Germany, on 8 May 1945. The *Panzerknacker* was soon surrounded by American airmen. In the background can be seen a parade of Lockheed P-38 Lightning twin-engined fighter aircraft. (*Archiv von Lutz*)